SACRAMENTO PUBLIC LIBRARY
828 "I" Street
Sacramento, CA 95814
09/21

BACKYARD BIRDS

Blue Jays

by Elizabeth Neuenfeldt

BLASTOFF! READERS

BELLWETHER MEDIA • MINNEAPOLIS, MN

Blastoff! Readers are carefully developed by literacy experts to build reading stamina and move students toward fluency by combining standards-based content with developmentally appropriate text.

 Level 1 provides the most support through repetition of high-frequency words, light text, predictable sentence patterns, and strong visual support.

 Level 2 offers early readers a bit more challenge through varied sentences, increased text load, and text-supportive special features.

 Level 3 advances early-fluent readers toward fluency through increased text load, less reliance on photos, advancing concepts, longer sentences, and more complex special features.

★ **Blastoff! Universe**

Reading Level: Grade K → Grades 1–3 → Grade 4

This edition first published in 2022 by Bellwether Media, Inc.

No part of this publication may be reproduced in whole or in part without written permission of the publisher. For information regarding permission, write to Bellwether Media, Inc., Attention: Permissions Department, 6012 Blue Circle Drive, Minnetonka, MN 55343.

Library of Congress Cataloging-in-Publication Data

Names: Neuenfeldt, Elizabeth, author.
Title: Blue jays / Elizabeth Neuenfeldt.
Description: Minneapolis, MN : Bellwether Media, 2022. | Series: Blastoff! readers : Backyard birds | Includes bibliographical references and index. | Audience: Ages 5-8 | Audience: Grades K-1 | Summary: "Developed by literacy experts for students in kindergarten through grade three, this book introduces blue jays to young readers through leveled text and related photos"– Provided by publisher.
Identifiers: LCCN 2021000672 (print) | LCCN 2021000673 (ebook) | ISBN 9781644874905 (library binding) | ISBN 9781648343988 (ebook)
Subjects: LCSH: Blue jay–Juvenile literature.
Classification: LCC QL696.P2367 N48 2022 (print) | LCC QL696.P2367 (ebook) | DDC 598.8/64–dc23
LC record available at https://lccn.loc.gov/2021000672
LC ebook record available at https://lccn.loc.gov/2021000673

Text copyright © 2022 by Bellwether Media, Inc. BLASTOFF! READERS and associated logos are trademarks and/or registered trademarks of Bellwether Media, Inc.

Editor: Betsy Rathburn   Designer: Andrea Schneider

Printed in the United States of America, North Mankato, MN.

Table of Contents

What Are Blue Jays?	4
Forest Friends	8
Flying in Flocks	14
Glossary	22
To Learn More	23
Index	24

What Are Blue Jays?

Blue jays are big **songbirds**. They are in the crow family.

All in the Family

Steller's jay

black-billed magpie

American crow

Blue jays are easy to spot. Their blue feathers stand out!

feathers

Forest Friends

Blue jays like to live near forests. They build round nests in trees.

nest

Blue jays eat seeds and nuts. They crack open hard shells with their **bills**.

Blue jays eat fruit and **insects**, too. They store extra food in **caches**.

Blue Jay Food

seeds fruit insects

Flying in Flocks

Blue jays form lifelong pairs. Pairs may live together in **flocks**.

In winter, some flocks **migrate** to find food. Others stay put. They eat stored food.

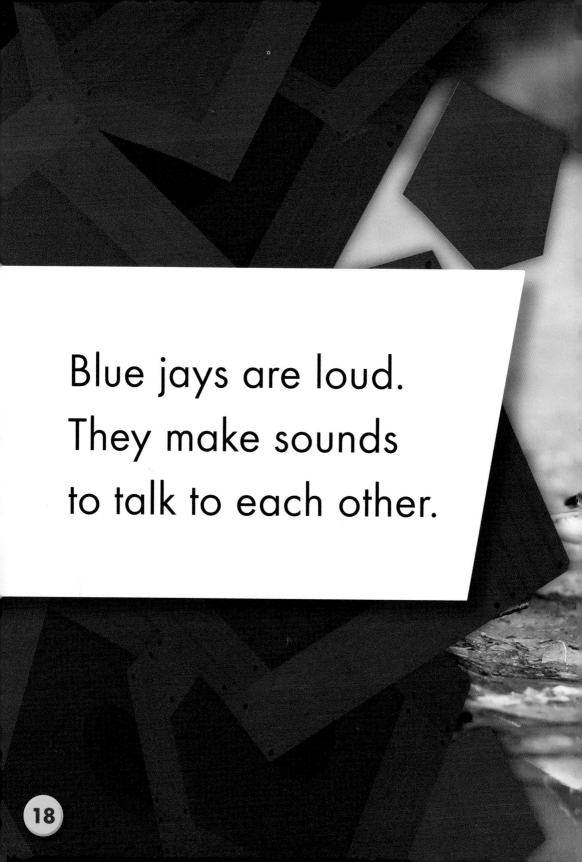

Blue jays are loud. They make sounds to talk to each other.

Blue jays can copy calls from other birds. They are smart birds!

Glossary

bills

the mouths of birds

insects

small animals with six legs and hard outer bodies

caches

places for hiding or storing food

migrate

to travel with the seasons

flocks

groups of birds

songbirds

birds that make musical sounds

To Learn More

AT THE LIBRARY

Neuenfeldt, Elizabeth. *Cardinals*. Minneapolis, Minn.: Bellwether Media, 2022.

Sewell, Matt. *The Atlas of Amazing Birds*. New York, N.Y.: Princeton Architectural Press, 2020.

Ward, Jennifer. *How to Find a Bird*. New York, N.Y.: Beach Lane Books, 2020.

ON THE WEB

FACTSURFER

Factsurfer.com gives you a safe, fun way to find more information.

1. Go to www.factsurfer.com.
2. Enter "blue jays" into the search box and click 🔍.
3. Select your book cover to see a list of related content.

Index

bills, 10, 11
caches, 12
calls, 19, 20
color, 6
family, 4, 5
feathers, 6, 7
flocks, 14, 15, 16
food, 12, 13, 16
forests, 8
fruit, 12
insects, 12
migrate, 16
nests, 8, 9
nuts, 10
pairs, 14, 15
seeds, 10

songbirds, 4
sounds, 18
trees, 8
winter, 16

The images in this book are reproduced through the courtesy of: Harold Stiver, front cover (blue jay); Artazum, front cover (background); FotoRequest, pp. 3, 7 (feathers), 22 (caches), 23; rob_knight_ink, pp. 4-5; teekaygee, p. 4 (Steller's jay); Nick Pecker, p. 4 (black-billed magpie); Melinda Fawver, p. 4 (American crow); Steve Brigman, pp. 6-7; Joseph M. Arseneau, pp. 8-9; David Tran Photo, p. 9 (nest); Rohane Hamilton, pp. 10-11; BirdImages, pp. 12-13; Tuzemka, p. 13 (fruit); SakSa, p. 13 (seeds); David Havel, p. 13 (insects); Rejean Bedard, pp. 14-15; blickwinkel/ Alamy, pp. 15 (flock), 18-19, 22 (flocks); Gary W. Carter/ Alamy, pp. 16-17; Steve Byland, pp. 20-21; RonnieWilson, p. 22 (bills); Marko Rupena, p. 22 (insects); Natalie Sexton, p. 22 (migrate); RobDun, p. 22 (songbirds).